THE MAYA PETÉN STOVE

Design and construction of a wood cooking stove

First Edition

March 2014

Oliver Style

IT CA

Appropriate Technology

The Maya Petén Stove

Design and construction of a wood cooking stove

First Edition.

ITACA Appropriate Technology Publications

Barcelona, Spain.

March 2014.

ISBN: 978-84-616-8640-7

Layout: Richard Grove

ITACA
Appropriate Technology

Dedicated to Silvestre Sales and the Salud Ambiental Team-Maya Petén in Guatemala, and Carles Yépez, Silvia Quilumbango and Gustavo León in Intag, Ecuador.

My thanks to Marianne Loewe, Dennis Garvey, John Straw, Cat Quinn, Richard Grove, Carles Bargalló, and the Concern America team, for their support over the years.

This publication has been possible with the support of Concern America

INDEX

1 INTRODUCTION

Around 2600 million people in the world depend on biomass for cooking. Biomass refers to fuels of organic origin, such as wood logs, charcoal, forest residues and dry animal dung. Humans have been burning biomass since they first began controlling fire around 400,000 years ago. Today, it constitutes around 7 % of world energy consumption.

Biomass offers an accessible and low-cost fuel source compared with alternatives such as gas or electricity. With sustainable forest management and processing, biomass can be a renewable energy source for cooking. However, it's important that the combustion process within the stove is efficient so that it uses a minimum quantity of fuel. An efficient stove offers the following advantages:

- ✓ Reduces the amount of time needed to source firewood

- ✓ Reduces the amount of toxic smoke in the kitchen

- ✓ Improves health conditions in the home

- ✓ Offers greater safety while cooking and reduces the risk of burns

- ✓ Reduces the effects of deforestation

This manual is a basic guide to the design and construction of a "Maya Petén" wood cooking stove. The design is the result of over 10 years work by the Maya Petén *Salud Ambiental* team in Guatemala, with the NGO Concern America. It was subsequently implemented in the Intag region of Ecuador. In an evaluation done by the project in 2011, it was found the stoves saved around 35 % fuel compared with traditional open fire cooking.

Here you'll find the following information:

- ➢ **Chapter 1** introduces the general context of cooking with biomass, project preparation and the outline design of the Maya Petén stove.

- ➢ **Chapter 2** deals with the principals of heat transfer in the stove and the most appropriate materials for the different components.

- ➢ **Chapter 3** presents the stove components, materials and cost of the Maya Petén stove.

- ➢ **Chapter 4** presents the design concepts and construction.

- ➢ **Chapter 5** contains information regarding operation, maintenance and shows you how to do a thermal efficiency test.

- ➢ **Chapter 6** includes a short bibliography if you want to dig deeper.

1.1 BIOMASS AND A GROWING POPULATION

When we began burning wood for cooking, there were a lot fewer of us around and in those days there were forests all over the place. Since then, world population has sky rocketed and there's much less forest around. Figure 1 shows the extent world population growth since the year 1000.

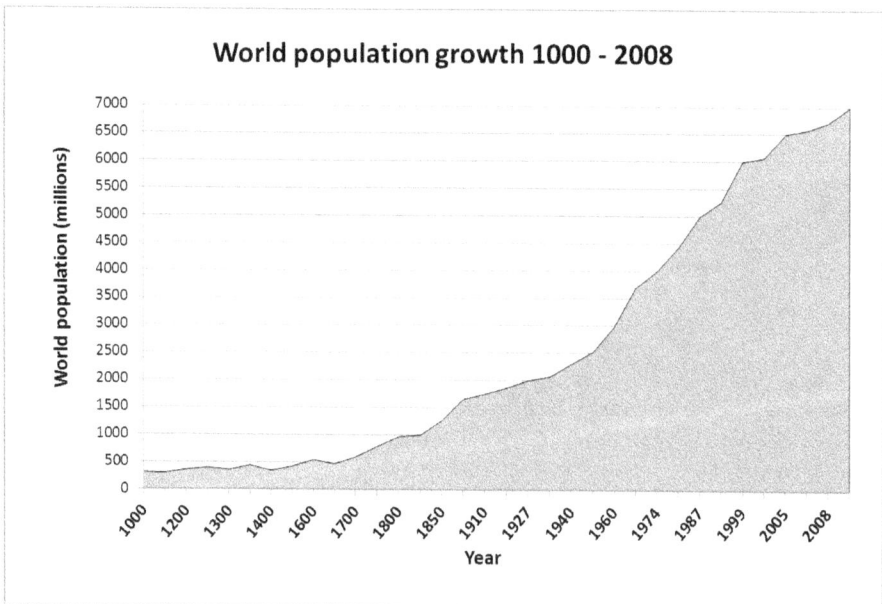

Figure 1: World population growth between 1000 and 2008

So, every year there´s more of us needing more and more fuel to cook. Table 1 shows the estimated number of people who currently use biomass for cooking and projections for the future, based on data from the International Energy Agency (IEA).

	2004 (millions)	2015 (millions)	2030 (millions)
Africa	579	632	725
Asia	1865	1922	1917
Latin America	83	86	85
TOTAL	2528	2640	2727
% increase over 2004 baseline	-	4 %	8 %

Table 1: Estimated number of people dependent on biomass for cooking in 2004 and future projections

The main cause of deforestation in developing countries continues to be large scale intensive agriculture, livestock grazing and timber exploitation. However, the burning of wood, coal, and forest residues for cooking has a negative environmental impact and adds to a problem that is now reaching crisis point in many parts of the world.

Figure 2: The 3-stone open fire, Petén, Guatemala (courtesy of Silvestre Sales)

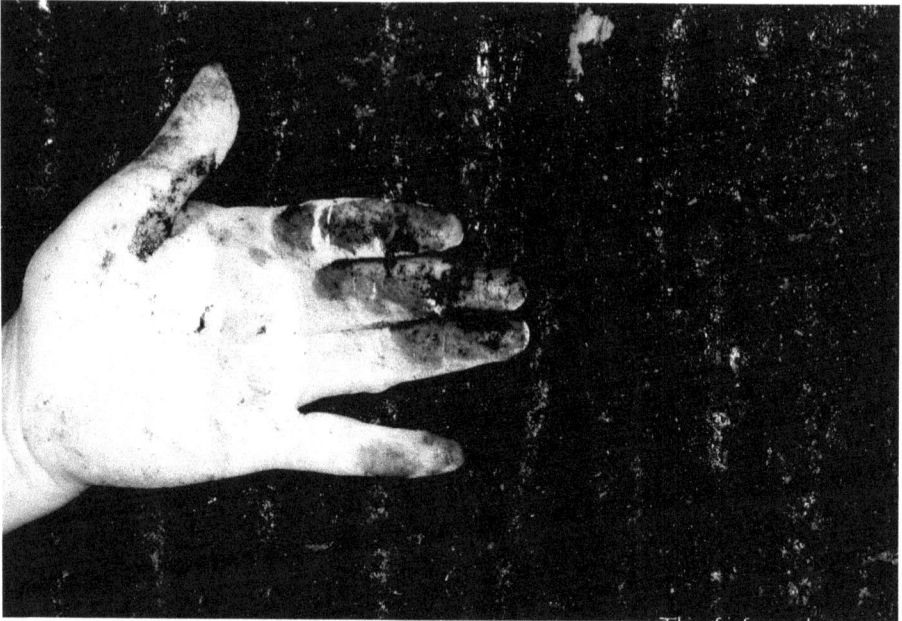

Figure 3: Soot deposits on the roof above an open fire, Petén, Guatemala (courtesy of Silvestre Sales)

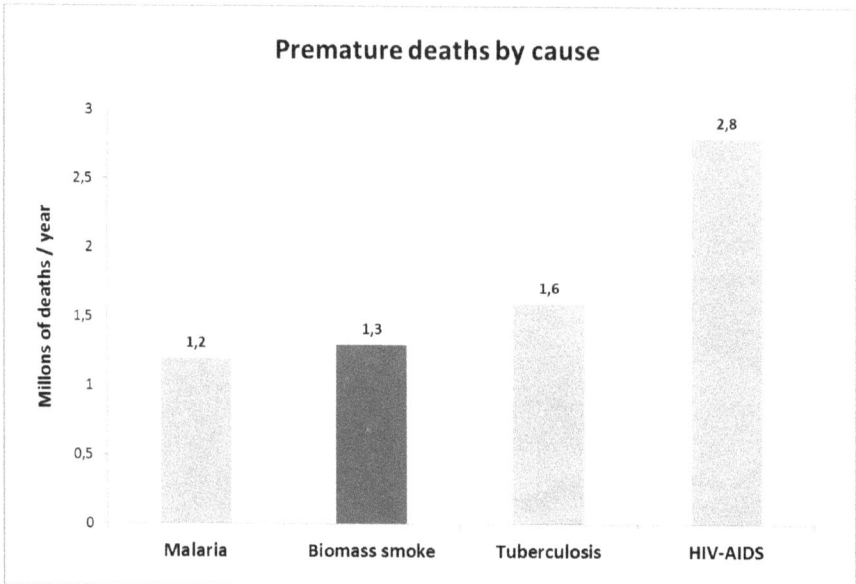

Figure 4: Annual premature deaths by cause

1.2   TAKE A DEEP BREATH...

In 2010, a study was done as part of the RESPIRE program (Randomized Exposure Study of Pollution Indoors and Respiratory Effects) on the effects of contaminant gases produced by the burning of biomass for cooking in developing countries. The study showed that the amount of smoke inhaled by a baby near an open fire is equivalent to puffing on 3 - 5 cigarettes a day. Figure 2 shows a traditional 3-stone open fire: quick and cheap to set up but one that creates a lot of smoke in the kitchen and is uncomfortable to use for whoever's cooking. Figure 3 shows the soot deposits on the roof above an open fire: this all ends up in the lungs of whoever is in the kitchen of course...

Burning biomass on an open fire produces smoke containing a large number of toxic substances, including carbon monoxide, formaldehyde, dioxins, and volatile organic compounds, that are carcinogenic and generally not very nice. These toxins cause respiratory diseases such as pneumonia and bronchitis, causing infant mortality and low birth weight. The IEA estimates that wood smoke in the home is responsible for the premature death of around 1.3 million people each year (Figure 4), more than both malaria and HIV-AIDS. That's quite of lot of people isn't it?

1.3   SCHEMATIC OF THE MAYA PETÉN STOVE

Figure 5 shows a schematic diagram of the Maya Petén stove. The design is a variant of the "Doña Justa" stove, the Aprovecho stove, the Rocket stove, and the ONIL "plancha" stove, based on work by Ben Dana @ AIDG and the Aprovecho Research Center. It consists of the following:

✓   **Outer walls:** these provide the stove base and body. They can be built from a variety of materials, including cement blocks, bricks, adobe or compacted earth bricks.

✓   **Fire box:** this is where the wood is burnt, channelling heat to the hotplate where food is prepared. The firebox is surrounded by insulating material to reduce heat loss to the outer walls. To allow for the entrance of sufficient fresh air and efficient combustion, a grate is placed on the floor of the firebox, which raises the logs off the floor, facilitating air flow.

✓   **Hotplate:** sheet metal that conducts heat from the fire to the food.

✓   **Chimney:** carries the flue gases out of the kitchen.

✓   **Ash hatch:** allows for ash and soot removal when cleaning the chimney.

Figure 6 shows a completed Maya Petén stove with a deluxe tile finish for easy cleaning.

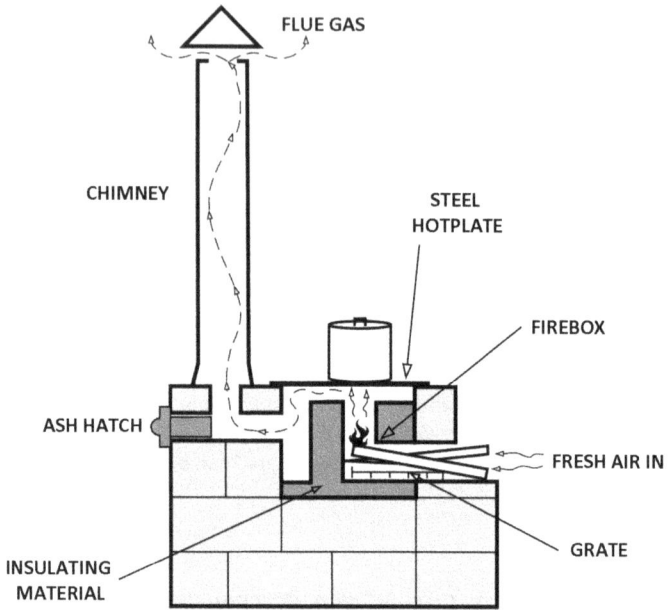

Figure 5: Schematic diagram of the Maya Petén stove

Figure 6: A completed Maya Petén stove, Intag, Guatemala

1.4 PLANNING

Here are some general pointers for building a wood cooking stove:

➤ **Simplicity, flexibility and local know-how:** this manual presents one kind of stove design, which has been adapted and developed to meet the needs of people living in rural areas of Guatemala. In other zones, you will need to adapt the design to suit local habits. Try and involve the end-users in the design process so that it has the greatest chance of acceptance. Follow the basic design principles shown here and find the simplest, cheapest, and most robust solution possible, using local know-how and skills.

➤ **Planning:** take time to plan your project and it´ll have more chances of success. If you´re going to be building a large number of stoves, call meetings with the construction team and end-users to agree on a work plan. For larger volumes, prefabrication and mass production off-site is the best solution, reducing costs and improving quality.

➤ **Training:** end-users need to receive some basic training on how to use and maintain the stove, so that it has a long useful life. Factor in the hand-over process and training when you are planning your project.

➤ **Evaluation and thermal testing:** make sure you do an evaluation a year after the stoves have been built. Do a thermal efficiency test following construction and again with the evaluation, 1 year later.

Figure 7 shows a flow diagram to help you with project planning.

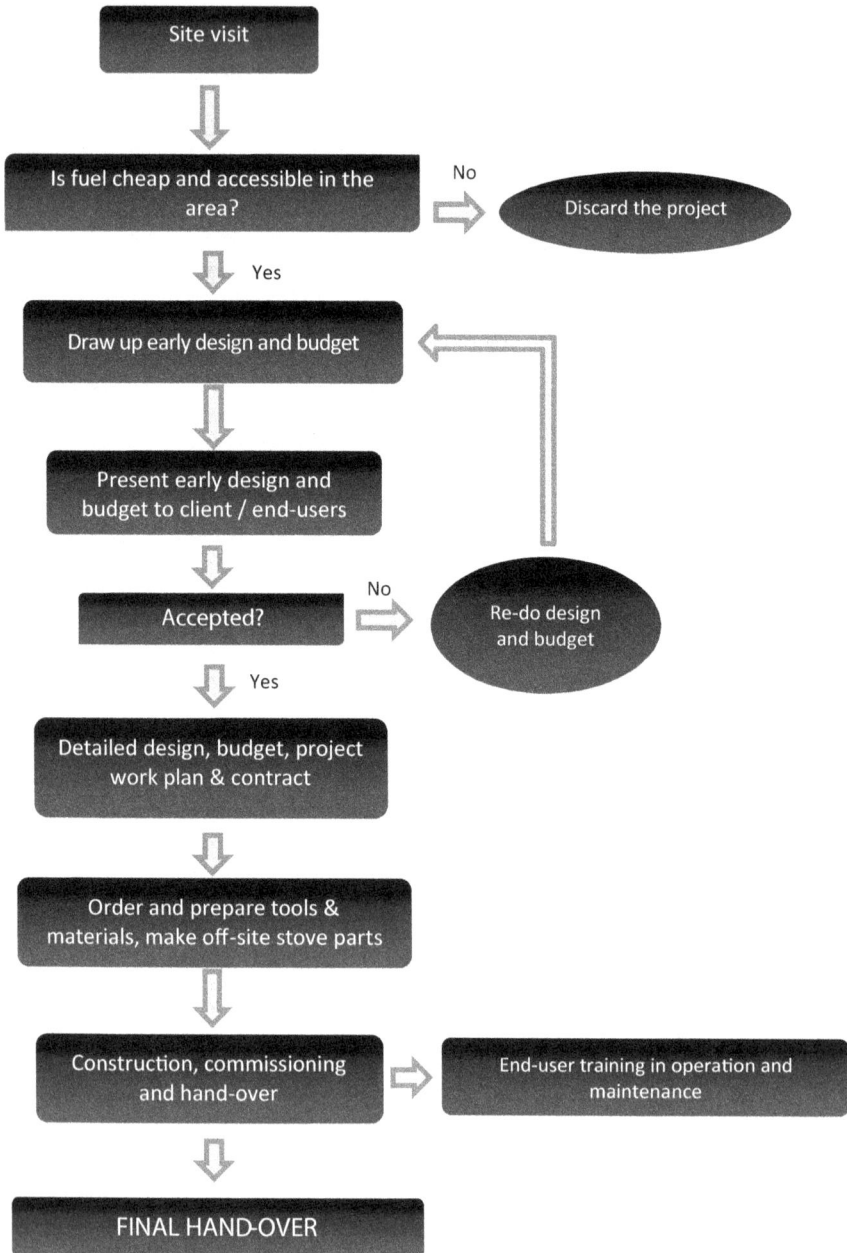

Figure 7: Flow diagram for carrying out a stove project

2 MATERIALS AND HEAT TRANSFER

Before getting into the design of the Maya Petén stove, let´s warm up with a look at some of the basic principles of heat transfer and the thermal properties of the materials we´re going to use in the stove. A basic understanding of the concepts presented here will help you tweak your design.

In theory, you need 18 grams of wood to cook 1 kg of rice. In practice, you need around 160 grams of wood to cook 1 kg of rice. Why 9 times more? The reason is that much of the heat is "Lost" as you cook the rice. There are three basic principles at work here:

> **Conduction and thermal inertia:** conduction refers to heat transfer between two materials, or through one material. This relates to the property of a material due to its thermal inertia, or its ability to absorb, store, and release heat.

> **Convection:** heat transfer through a fluid (this can be air or water).

> **Radiation:** heat transfer in the form of electromagnetic waves.

Evaporation is also a factor. Let's have a look at each one in turn.

2.1 CONDUCTION AND THERMAL INERTIA

If I light a fire and put a metal pot above the flames, the particles in the pot that are in contact with the fire, start to vibrate. Gradually this vibration extends to the rest of the pot and its temperature rises. This is due to *conduction* heat transfer. There will become a point where you will burn yourself if you pick up the lid without a cloth of some sort, as the pot will have *conducted* a large amount of heat from the fire to the lid handle (Figure 8).

Figure 8: Conduction heat transfer...ouch

If the pot was made of clay or ceramics, the temperature of the lid would be lower and you´d probably be able to pick up the lid without burning yourself. This is because clay is less conductive than metal. Different materials have different levels of thermal conductivity.

This is important in our stove, because it means we can select different materials that are best adapted to what we need: we can reduce heat loss towards the outer walls by insulating around the firebox with a material that has a low thermal conductivity (such as ash, pumice or vermiculite); and we can maximise heat transfer from the fire to the pot by using a material that has a high thermal conductivity (such as steel).

We can measure conduction heat transfer if we know the thermal conductivity of a material (alongside various other parameters that we won´t go into here...). Table 2 shows the conductivity of different materials that are typically used in a Maya Petén stove.

Related to conduction heat transfer is the thermal inertia properties of a material, or it´s heat capacity. A material with high thermal inertia, when exposed to a source of heat, acts like a water tank that´s being filled with water: it can store it for a time, and then release it when it´s needed. A similar thing happens with heat: a material can absorb heat and release it cyclically. A material's thermal inertia or heat capacity depends on its density, specific heat capacity and thermal conductivity. Table 2 shows these three properties for materials that are typically used in a Maya Petén stove.

If I build a stove and pack earth around the firebox, the earth will absorb and store quite a lot of heat from the fire. This means that if my stove is cold when I light it in the morning, I'll need to burn quite a lot of wood before being able to cook properly, due the thermal inertia of the earth which absorbs heat and makes my stove heats up more slowly than it otherwise would. However, if I insulate around the firebox with an insulating material such as ash, pumice or vermiculite, my stove will take much less time to heat up, I'll use less fuel, and I'll be able to have a cup of coffee in much less time, which is always a good thing in the morning.

A frequent complaint from people who are used to cooking on open fires and who start using a stove such as the Maya Petén, is that it takes a lot of time to heat up. This happens because the conductivity and thermal inertia of the materials used in the stove are not taken into account properly in the design.

How can I best manage conduction heat transfer and thermal inertia in my stove?

➢ To get heat from the fire to the food I'm cooking, I'll use a material that is highly conductive, such as steel. Steel transfers heat fairly quickly and it's also strong, so it works well as a hotplate.

➢ On the other hand, around the firebox I want to make sure that as little heat as possible escapes towards the outer walls, so I'll need a material that isn't conductive,. A material that insulates well is light and full of air pockets, and acts as a big woolly jumper wrapped around the firebox, keeping the heat in. The firebox should not be in contact with earth, stone, sand, gravel, or concrete, as these conduct and absorb heat rather than insulating.

➢ It's a good idea to encourage end-users to cook with metal crockery rather than clay or ceramic pots and pans, as metal conducts heat more effectively.

MATERIAL	THERMAL CONDUCTIVITY λ (W/m.K)	DENSITY (kg/m³)	SPECIFIC HEAT (J/kg.K)
Aluminium	230.0	2700	880
Steel	45.0	7800	450
Concrete	2.0	1920	1000
Earth	1.3	1400	1000
Adobe blocks	1.1	1800	880
Cement or lime mortar	1.0	1890	1000
Bricks	0.9	2400	840
Ceramic floor tiles	0.8	1800	800
Concrete blocks	0.5	1460	1000
Gravel	0.4	1840	840
Ash	0.2	400	650
Pumice, vermiculite	0.1	200	1000

Table 2: Thermal conductivity, density and specific heat of different materials used to construct a stove

2.2 CONVECTION

If I take the pot off the fire and put my hand a certain distance above the flames, I'll be able to feel convection heat transfer. The air particles just above the flames start to vibrate, they become less dense compared to the cooler surrounding air, and they rise (Figure 9). If I keep my hand there for long enough it'll soon start to look like a well-cooked hamburger and it might become a little uncomfortable.

Figure 9: Convection heat transfer…uyy

Convection heat transfer is the most important mode of heat transfer in the stove, since this is the main way in which heat is moved from one place to another: via hot air, or via the flue gases. The open fire we saw in Figure 2 is relatively inefficient because the convection heat transfer is not controlled very well: some heat reaches the tortillas but a lot of it goes elsewhere.

In the Maya Petén stove shown in Figure 5, convection heat transfer is controlled to a greater extent: cool fresh air comes in to the firebox, is heated and then brushes past the pot, continuing on below the hotplate before leaving through the chimney.

How can I best control convection heat transfer in my stove?

➢ I need to make sure, as far as possible, that the cross sectional area from the mouth of the firebox to the exit through the chimney is constant. This means the flue gases will have a more constant velocity as they move through the stove. A constant velocity through the same cross sectional area will mean a constant flow rate, meaning less turbulence and reduced energy losses.

➢ The other thing I can do is have a special hotplate made that allows pots of a given diameter to sink down into the air space below. This will provide more pot surface area for heat transfer to take place, so that the hot flue gases can brush over a greater surface area of my pot. The drawback to this is that it adds cost, complication and means end-users can only use one size pot. The Maya Petén design sacrifices some efficiency here, with a hotplate that has a series of concentric rings cut out. Depending on the size of the pot or frying pan, rings can be added or removed, but pots and pans always sit on the hotplate, without dropping down below. This reduces convection heat transfer somewhat but is more practical and simple.

➢ It's a good idea to advise end-users to always have as many rings of the hotplate removed as possible, so that the maximum surface area of the bottom of the pot is available for convective heat transfer to take place.

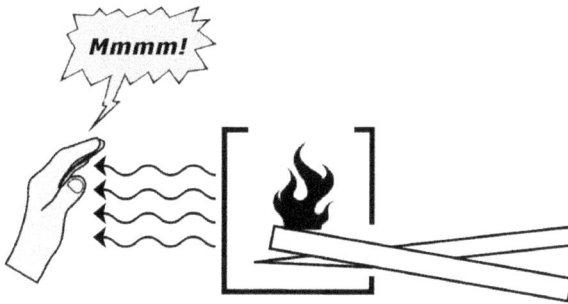

Figure 10: Radiation heat transfer...mmmm

2.3   RADIATION

Radiation refers to heat transfer in the form of electromagnetic waves that travel through the air (they can also travel through a vacuum, such as in space). So, for example, if I put my hand to one side of a fire, I can feel heat transfer through radiation. The flames of the fire make the electrons in the air vibrate and this generates an electromagnetic wave that arrives at my hand. I feel this energy in the form of heat (Figure 10). This is the same reason we feel warm when sit outside in the sun with no clothes on.

How can I best manage radiation heat transfer in my stove?

- ➢ I'll tell end-users to make sure that the rings on the hotplate are shut when not cooking, so that radiating heat isn't "lost" to the ceiling of the kitchen.

- ➢ I'll advise end-users to always have the as many rings of the hotplate removed as possible when cooking, so that the maximum surface area of the bottom of the pot can "see" the flames. This is referred to as the "view factor" (see Baldwin 1986 and FAO 1993), or the relative area of the emitting surface (the fire) in proportion to the absorbing surface (the bottom of the pot). The greater the view factor the greater radiative heat transfer can take place.

2.4   EVAPORATION

This phenomenon is not strictly to do with the stove design, but it's worth thinking about. Evaporation is a physical process which relates to a change of state between a liquid and a gas, what's known as a phase change. If I put a pot full of water over the fire, when the temperature of the water reaches a certain point (the boiling point, 100 ºC), the water particles will have enough energy to make the jump from a liquid state into a gas, or into

Figure 11: Evaporation...ssss

water vapour/steam (Figure 11). Energy from the fire that has been transferred into the pot and the water in it, is therefore taken away in the steam that leaves the pot.

How to I reduce evaporation heat loss in the stove?

➢ By advising end-users to keep the lids on all the pots and pans and reduce the amount of steam being lost.

2.5 OTHER CONSIDERATIONS

Logically, there are other factors at play here, beyond the thermal interactions mentioned above, which will influence the materials I choose for building the stove. We need materials that are:

➢ **Durable:** sufficiently strong to deal with the heat and the knocks that a stove will go through in day-to-day use.

➢ **Low-cost:** the stove needs to be economically accessible to people.

➢ **Safe:** it's important the materials we use in a stove don't emit toxic substances when they heat up.

➢ **Workable:** we need to find materials that are relatively malleable and easy to work with, so that we can build the stove quickly and easily.

3 COMPONENTS, MATERIALS AND COSTS

This chapter describes the components and materials of the Maya Petén stove. If you´re not able to find the materials presented here in your area, find alternatives that meet the required thermal and mechanical properties described above in Chapter 2.

Figure 12: A Maya Petén stove with extended base, Intag, Ecuador

Figure 13: Cement blocks and bricks

Figure 14: Compacted earth blocks

3.1 OUTER WALLS

The outer walls provide the structural shell of the stove and the base on which the hotplate and chimney are mounted. The most adequate materials are:

- ➤ Cement blocks

- ➤ Clay fired bricks

- ➤ Abode or compacted earth blocks

- ➤ Stone masonry

Examples of these materials can be found in Figure 13, Figure 14, and Figure 15. Due to their relatively low cost, durability and availability in most areas, concrete cinder blocks are usually the best option for building the outer walls. These can also be in the form of pre-fabricated concrete components that interlock to provide the required form and structure.

3.2 FIREBOX

The firebox is the combustion engine of the stove, providing a contained space in which the logs are burned, channelling the hot flue gases to the hotplate where food is being prepared. Fresh cool air is drawn into the firebox through its mouth and is heated as it passes the fire, rising vertically towards the hotplate and transferring heat to the cooking pot. Thick ceramic floor tiles are the most suitable low-cost material for the firebox. They have a relatively low thermal conductivity (λ = 0.8 W/m.K), can support the high temperatures of the firebox ($\approx$ 200 to 500 ºC), and survive years of knocks from logs being inserted (Figure 16). Be careful not to use tiles that are too thin as they will quickly break. They are easy to cut with an angle grinder (Figure 17).

Figure 15: Out walls made with concrete cinder blocks, Petén, Guatemala (courtesy of Silvestre Sales)

Figure 16: Firebox made from fired ceramic floor tiles

Figure 17: Cutting ceramic floor tiles for the firebox with an angle grinder

Figure 18: Insulating material around the firebox, Petén, Guatemala (courtesy of Silvestre Sales)

3.3 INSULATING MATERIAL

The insulating material is placed under and around the firebox, acting as a big woolly jumper that reduces heat loss to the outer walls and makes sure the majority of heat is transferred to the hotplate (Figure 18). The insulating material must have a low thermal conductivity, such as ash, pumice or vermiculite (see Table 2).

3.4 HOTPLATE

The hotplate transfers heat from the firebox to the food that's being prepared. The best material is steel, which has a high thermal conductivity (45 W/m.K). The recommended thickness is between 4 and 6 mm. Less than 4 mm and the plate won't last long; more than 6 mm and it'll take too long to heat up from a cold start. The dimensions of the hotplate and cooking rings can vary depending on the needs of end-users and available budget (Figure 19, Figure 20). Two-hole hotplates means food that has been cooked can be kept warm on the back ring. The smaller the hotplate, the lower the cost, and less heat loss your stove will have.

Figure 19: A completed Maya Petén stove, with a one-hole one-ring hotplate, Petén, Guatemala (courtesy of Silvestre Sales)

Figure 20: Two-hole, two-ring hotplate

Figure 21: Chimney pipe made from concrete and metal, Petén, Guatemala (courtesy of Silvestre Sales)

Figure 22: Chimney pipe made from concrete drainage pipes, Intag, Guatemala

3.5 CHIMNEY

The chimney carries the flue gases outside of the kitchen, improving health conditions for the people cooking. For durability and protection from burns, concrete drainage pipes are a good lasting solution (Figure 22). They can be combined with thin sheet metal pipes (Figure 21), although these tend to be shorter lasting.

3.6 COSTS

Table 3 shows the final materials list and Budget for the stove (Guatemala prices converted to USD$ @ 2012 prices):

Material	Quantity	Unit cost (US$)	Total cost (US$)	% of total
Sand (20 litre bucket)	4	$1,29	$5,15	4%
Cement (50 kg sack)	1	$9,65	$9,65	7%
Floor tiles	7	$0,64	$4,50	3%
Chimney pipes	2	$6,50	$13,00	9%
Cement blocks	38	$0,64	$24,45	17%
Steel hotplate	1	$41,17	$41,17	29%
Tools	1	$5,00	$5,00	4%
Transport	1	$6,43	$6,43	5%
Labour	2	$12,87	$25,73	18%
Miscellaneous	1	$5,79	$5,79	4%
TOTAL			$141	

Table 3: Materials budget

Once you've prepared the budget do a cost analysis. This'll help you see where you need to make savings and reduce the overall cost of the stove. Figure 23 shows a pie chart example, based on the costings above.

You can see that the largest material costs are the hotplate (29 %), cement blocks (17 %), and chimney pipes (9 %), constituting 55 % of the total cost of the stove. This is a specific example for a specific location: wherever you are, look for savings so that the stove is economically accessible to as many people as possible.

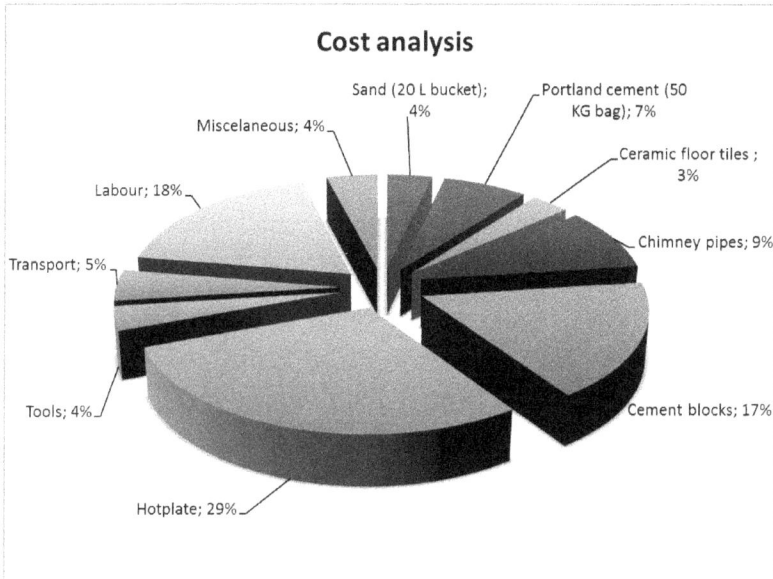

Figure 23: Cost analysis

4 DESIGN AND CONSTRUCTION

This chapter deals with the design and construction of the Maya Petén stove. If you understand the basic concepts presented here, you'll be able to adapt the design to suit local needs and materials, so that it works efficiently and has the greatest chance of being accepted by end-users. Above all, explain the design simply and visually to end-users before building anything, and ask for their feedback (Figure 24). Figure 25 shows a 3D render of the stove design presented here.

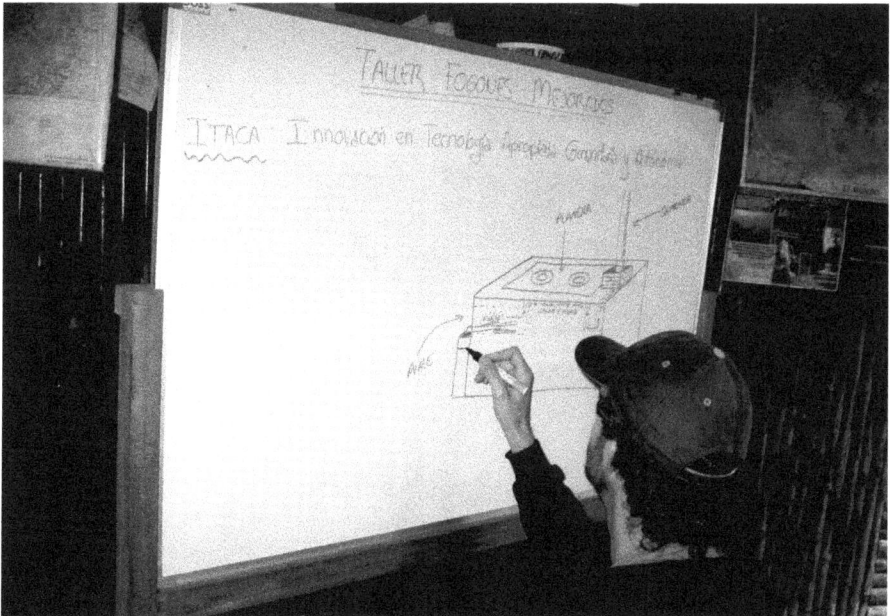

Figure 24: Explaining the design of the Maya Petén stove to end-users, Intag, Ecuador

Figure 25: 3D render of the Maya Petén stove

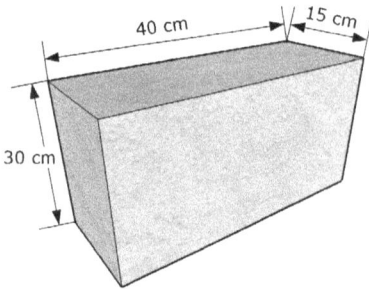

Figure 26: Typical block dimensions

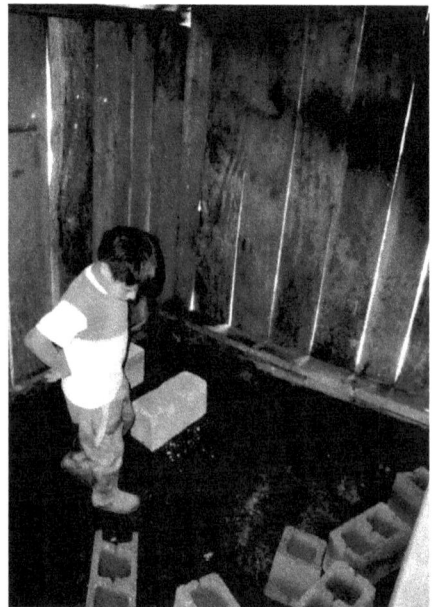

Figure 27: A construction manager checking the site levelling

4.1 OUTER WALLS, INFILL MATERIAL AND LOG LOADER

The dimensions of the outer walls will depend on the dimensions of the blocks you are going to use and the dimensions of the hotplate (Section 4.5). The example shown here is for blocks that measure 30 cm x 40 cm x 15 cm (height / length / width), shown in Figure 26. You can save on materials if you build the stove into the corner of room or against a wall. The first step in construction is to prepare the area where you're going to build the stove, making sure it's compacted down and perfectly levelled (Figure 27).

Outer and inner wall dimensions are shown in Figure 28 and Figure 29.

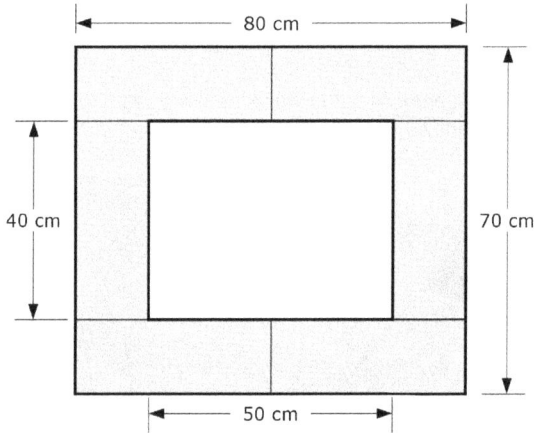

Figure 28: A construction manager checking the site levelling

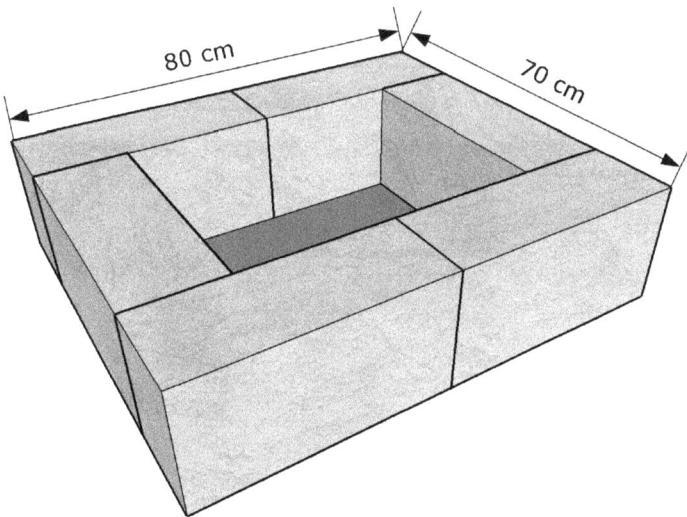

Figure 29: Outer and inner wall dimensions, plan view

With the second level of blocks you'll get up to around 40 cm in height (Figure 30). Only use a minimal amount of mortar between levels, and you can generally leave them dry between blocks on the same level (the outer wall structure doesn't need to be super strong, and this will save on mortar). You'll need to cut two blocks at ≈ 10 cm and another two at ≈ 30 cm (Figure 30). Cut the blocks with an angle grinder for clean cuts (Figure 31).

Once the first two levels of blocks are up, you can build the log loader and put in the in-fill material. The log loader is two blocks high with two extra blocks cut to a height of ≈ 6 cm (Figure 32). The in-fill material (earth or similar) is placed in the hole and lightly compacted.

Once the third level of blocks is complete, fill in with ≈ 4 cm of insulating material, to insulate under the firebox, shown in Figure 33 (logically this is for illustration purposes only, you first need to place the third level of blocks and *then* fill with insulating material).

On the third row, you'll need to cut two blocks at ≈ 8 cm (Figure 34).Things are hotting up! Let's move on to the firebox...

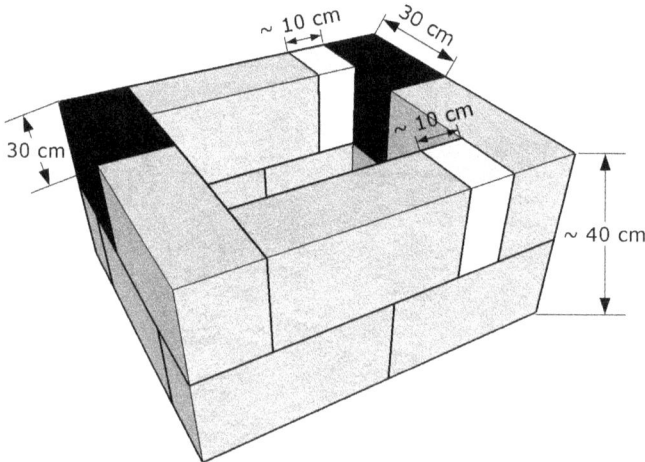

Figure 30: Outer wall dimensions, 3D view

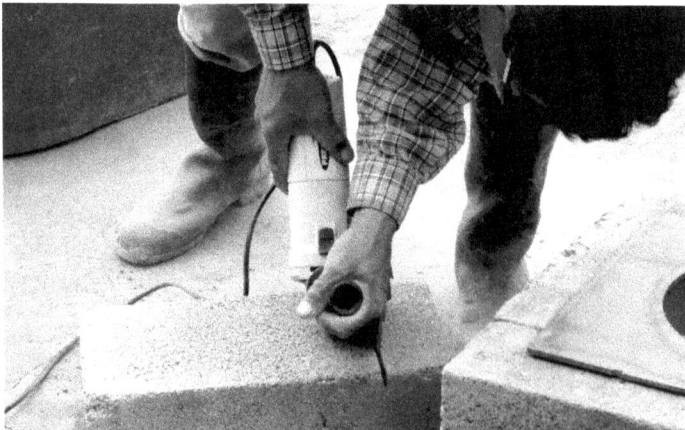

Figure 31: Cutting blocks with an angle grinder, Intag, Ecuador.

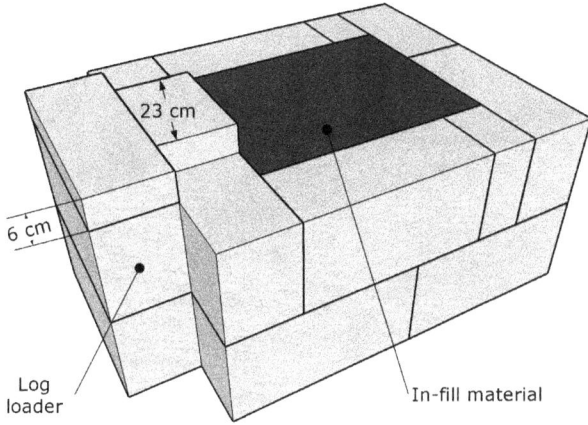

Figure 32: Second level of blocks, with cuts

23 cm

6 cm

Log
loader

In-fill material

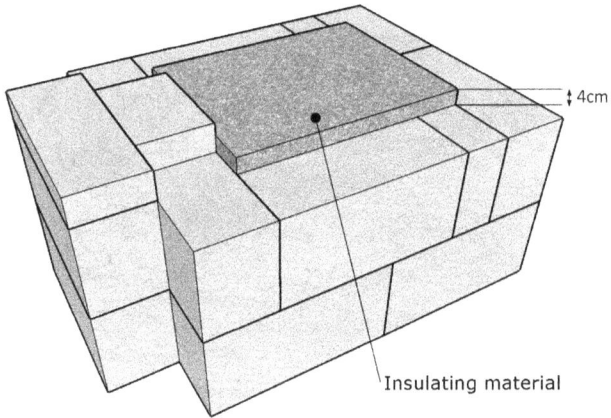

Figure 33: Insulating material beneath the firebox

4cm

Insulating material

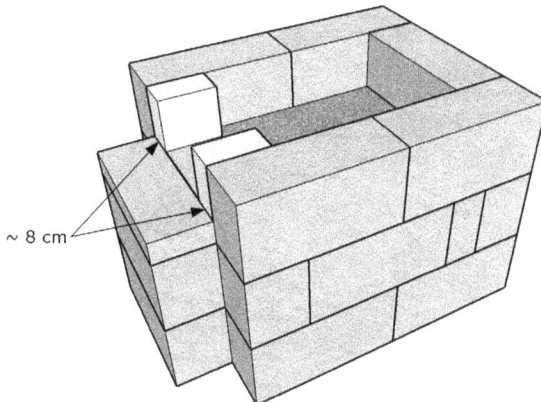

Figure 34: Third level of blocks

~ 8 cm

Figure 35:   Ceramic floor tile

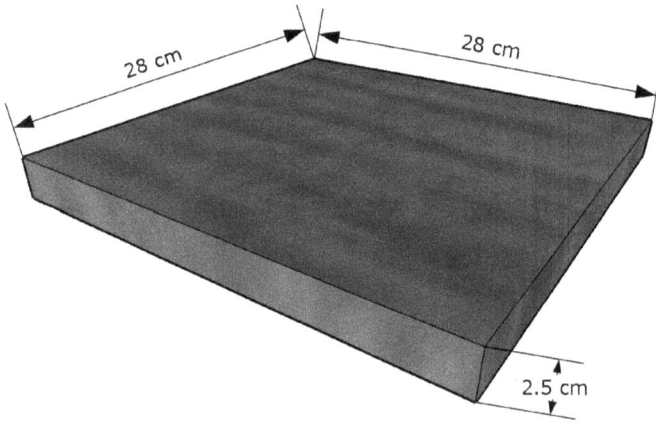

28 cm

28 cm

2.5 cm

Figure 36: Floor tile dimensions

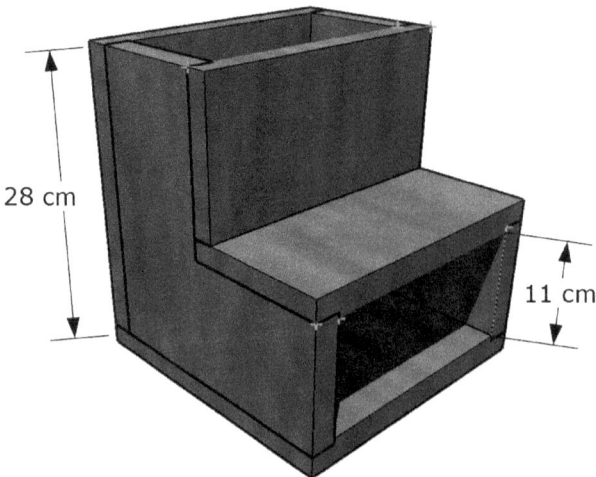

28 cm

11 cm

Figure 37:   Height ratio of ≈ 1 : 2.5 between mouth and tower

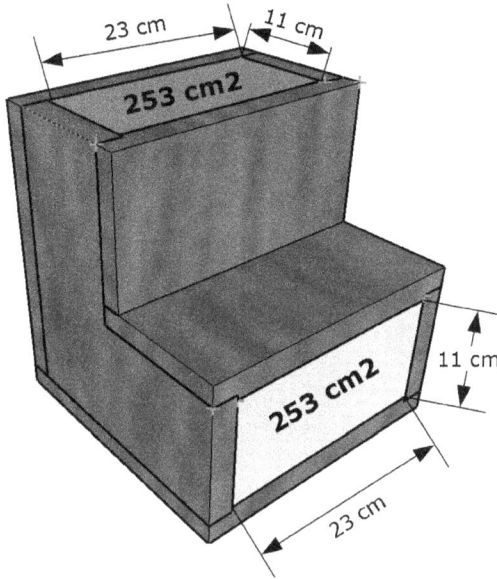

Figure 38: Identical cross sectional area of mouth and exit

4.2 FIREBOX

Ceramic floor tiles for the firebox can come in a variety of sizes. Have a look at what's available in your area. The firebox design presented here is for tiles that measure 28 cm x 28 cm x 2.5 cm (Figure 35, Figure 36). The dimensions of the tiles you use will influence the design of the firebox and pretty much everything else in the stove. Begin with the design of the firebox based on your tile dimensions, and work outwards from there, finding a solution that meets the design criteria presented here and which implies the least number of cuts and minimal generation of waste. Figures 41 to 45 show the dimensions and cuts you need to make for the firebox, based on the tile dimensions shown above.

4.2.1 Mouth-tower height ratio

The height ratio between the mouth of the firebox and the height of the tower needs to be approximately:

➤   1 : 2.5

This will help the stove draw properly and increase combustion efficiency. For example, for a mouth height of 10 cm, the tower needs to be around 25 cm high. For tiles with the dimensions shown in Figure 35, the best design to meet the mouth-tower height ratio is with 11 cm mouth height and 28 cm tower height (Figure 37). This reduces the number of cuts that need to be made and reduces waste.

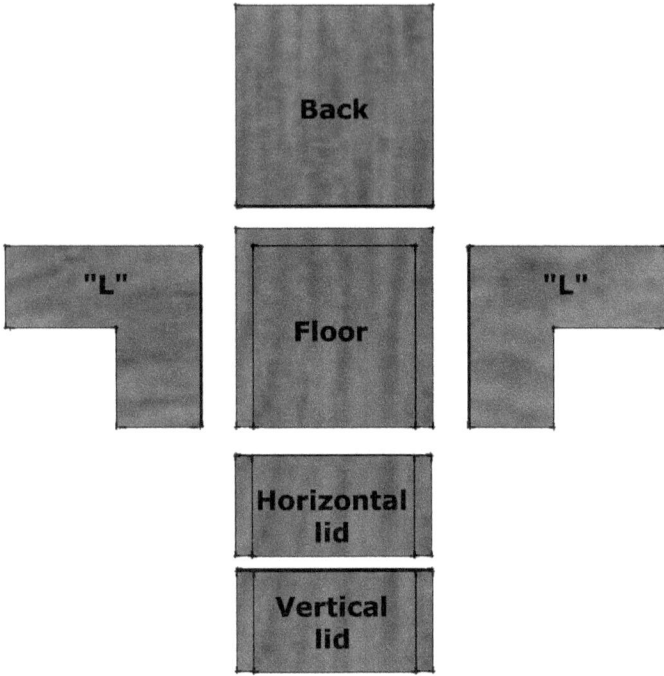

Figure 39: Firebox parts

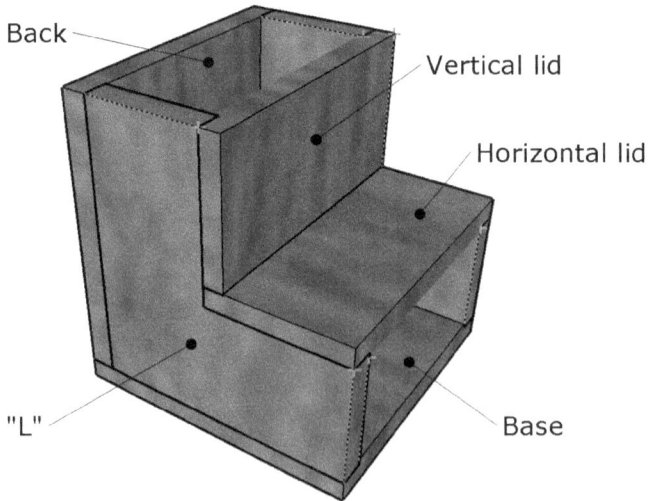

Figure 40: Firebox parts

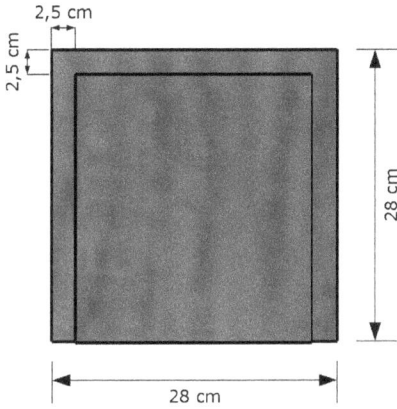

Figure 41: Dimensions and cutbacks Base

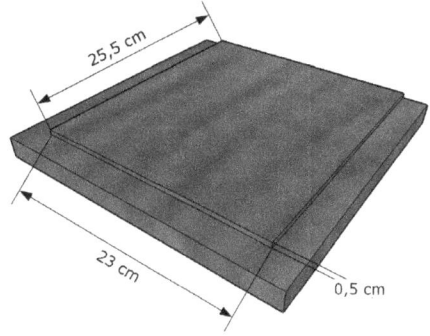

Figure 42: Base cutbacks

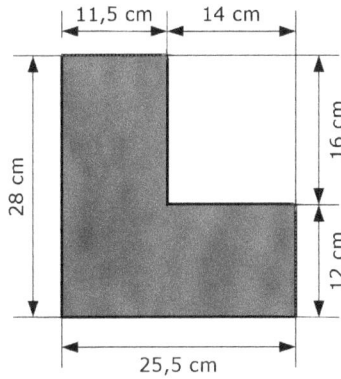

Figure 43: Dimensions and cuts "L"

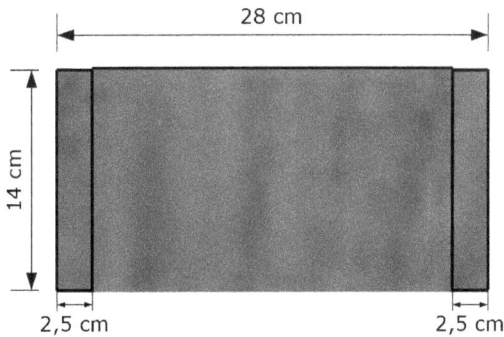

Figure 44: Dimensions and cuts Lids

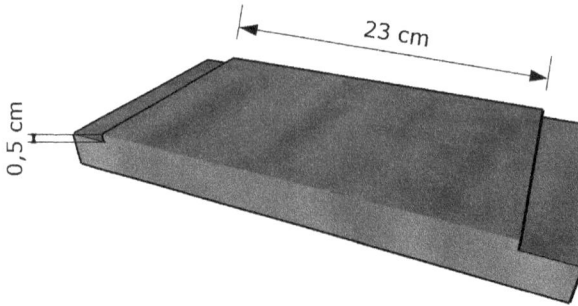

Figure 45: Lid cutbacks

4.2.2   Constant flue gas velocity

The stove's efficiency will be greater if the velocity of the flue gases remains fairly constant throughout the stove. Changes in the velocity of the gases will take place if the cross sectional area through which they flow varies, resulting in a varying flow rates. This in turn will result in energy losses, given that: ***Flow rate = fluid velocity x cross sectional area*** (see Dana 2009). This means we need to try and make sure the cross sectional through which air flows throughout the stove equal at all points, from the mouth of the firebox to the chimney. In terms of the firebox, this means the mouth and exit needs to have the same cross sectional area (shown in Figure 38).

4.2.3   Firebox parts

The firebox consists of 6 parts (Figure 39). The only part that doesn't need cutting is the Back. To meet the design criteria described above, the dimensions of each part are described below. The cutbacks allow all the parts to notch into one another and gives the firebox more strength.

The firebox parts can be sliced up and cut back with an angle grinder (Figure 46). Wet the tiles first and there'll be less dust when you cut. Once the parts are cut and ready, they can be assembled and the firebox can be placed in the hole (Figure 47, Figure 48).

To improve combustion in the firebox, have a grate made from rebar. This will lift the logs off the floor of the firebox and improve fresh air in-flow. The firebox is quickly filled up with ash and this blocks the air flow.

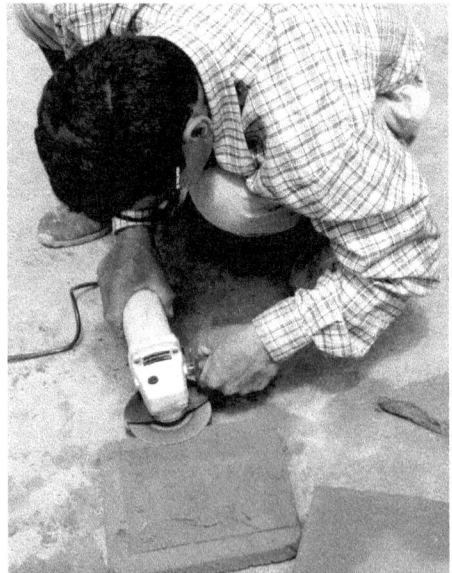

Figure 46: Making the cutbacks on floor tiles with an angle grinder

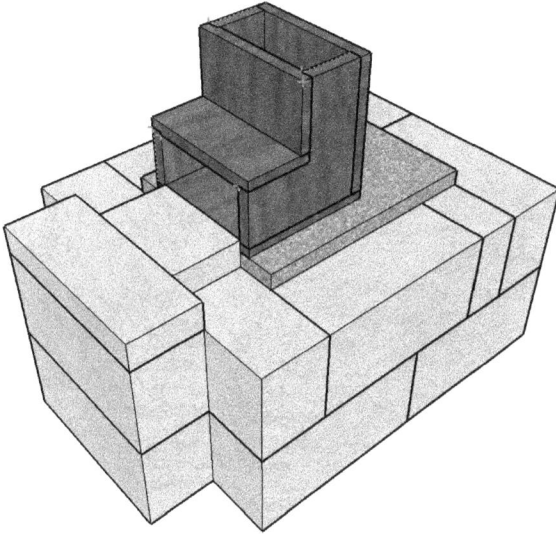

Figure 47: Firebox, second level

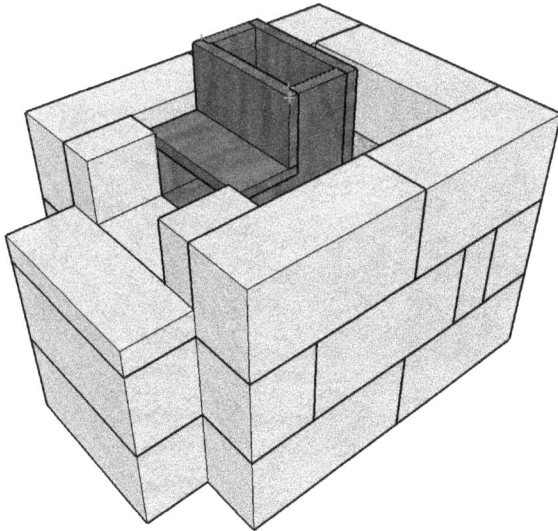

Figure 48: Firebox, third level

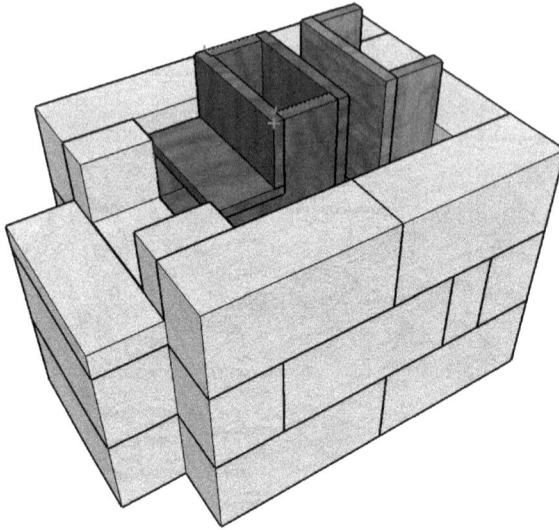

Figure 49: Firebox and exit

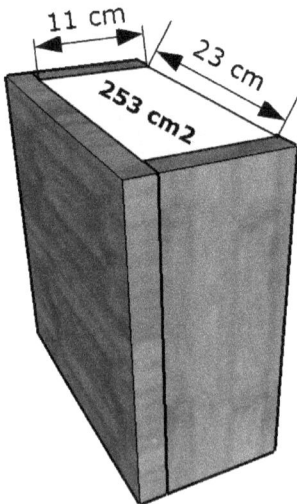

Figure 50: Dimensions and cross sectional
area of exit

4.3   EXIT

The exit takes the flue gases towards the chimney after travelling along under the hotplate. It consists of three tiles that form a channel where smoke can leave the stove and exit through the chimney.

To maintain a constant flow rate, the exit needs to have the same cross sectional area as the mouth and exit of the firebox. In this case, it is 253 cm^2 (Figure 50). The exit block needs to be cut to allow the flue gases through to the chimney, with the same cross sectional area as above (Figure 51).

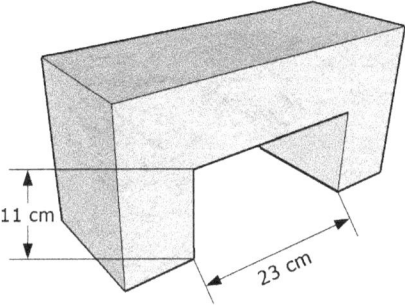

Figure 51: Cut in exit block

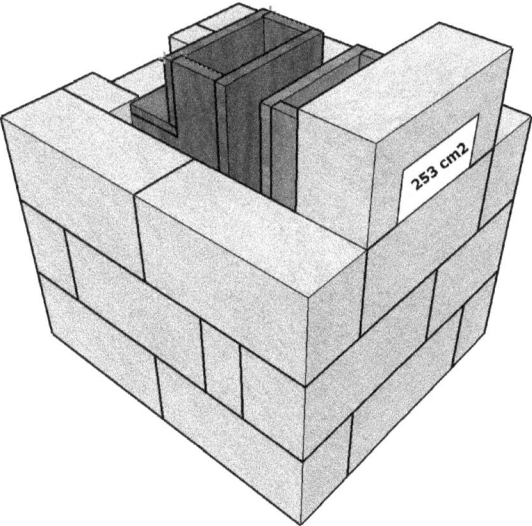

Figure 52: Exit block

Insulating material

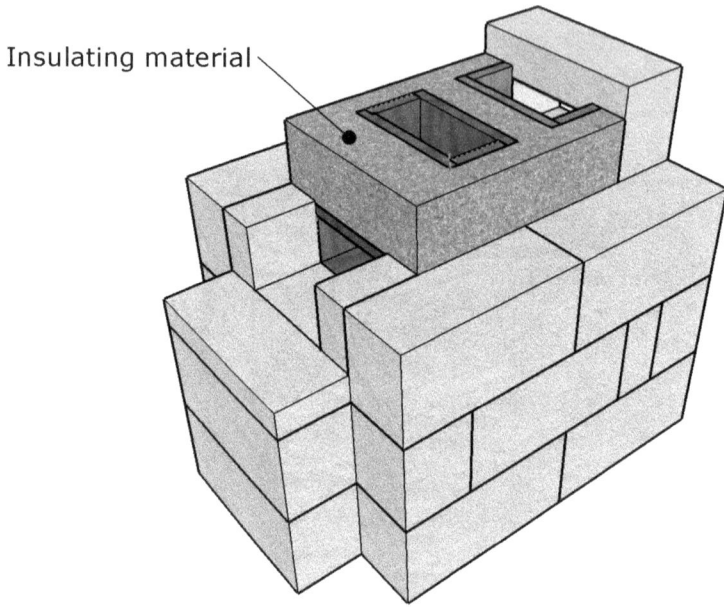

Figure 53: Insulating Material

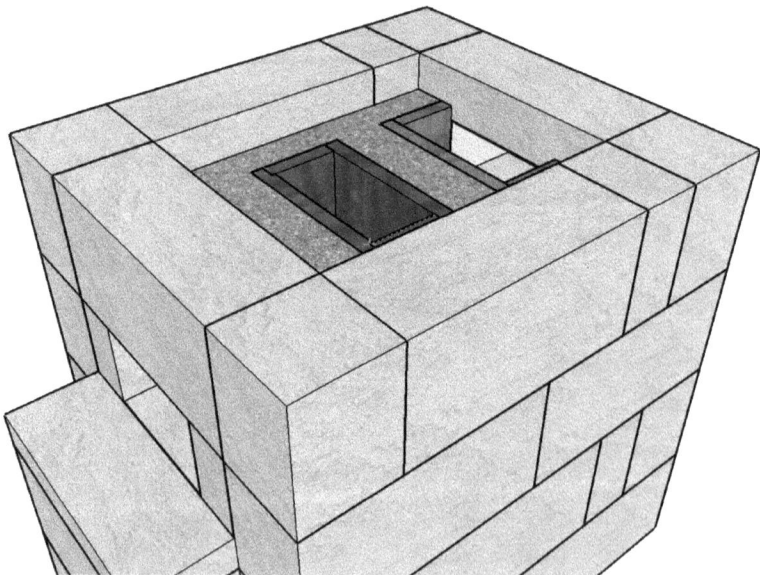

Figure 54: Fourth level of blocks with insulating material

4.4   INSULATING MATERIAL

To reduce heat loss to the outer walls, you need to insulate around the firebox, with ash, pumice or vermiculite (see Chapter 2). For illustrative purposes, the insulating material can be seen in Figure 53. Logically you'll need to lay the fourth level of blocks before filling in the insulating material (Figure 54).

To maintain a constant cross sectional area, the space beneath the hotplate needs to be approximately 6.5 cm high. For a width of 40 cm, this will provide 260 cm^2 of cross sectional area, when viewed from the front of the stove (very close the 253 cm^2 we have everywhere else). See Figure 55.

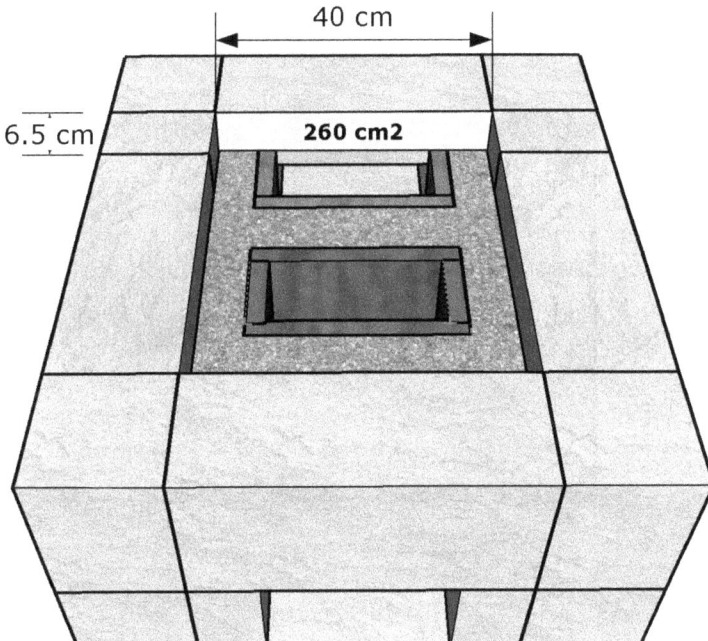

Figure 55: Cross sectional area below the hotplate

4.5 HOTPLATE

The dimensions of the hotplate and the number of holes and rings will vary. Remember that a smaller hotplate will be more efficient and the stove will be cheaper. The dimensions of the hotplate for the design presented here can be seen in Figure 56. It's important that the main hole is located write above the firebox exit (Figure 57), to maximise convective and radiant heat transfer.

Make sure a hole in the centre of the inner ring is included in the specification you send to the workshop where the plates are made, together with a hook for lifting the rings out when the stove is in use and hot (Figure 58).

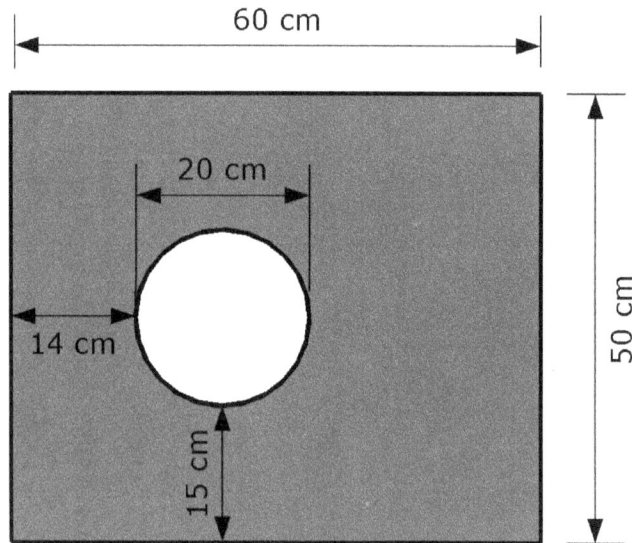

Figure 56: Hotplate dimensions

Figure 57: Location of main hole in relation to firebox exit

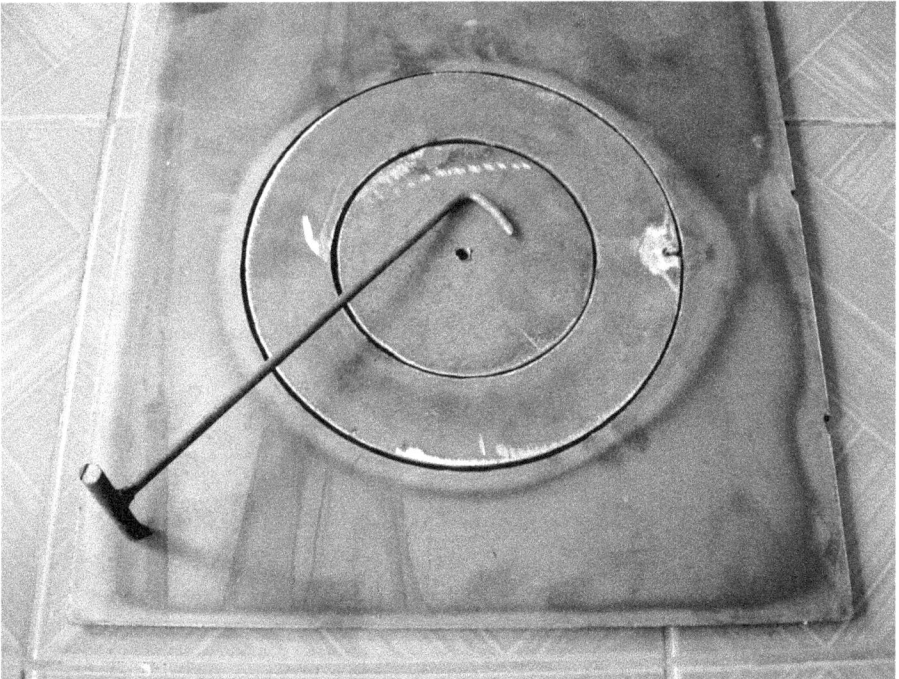

Figure 58: Hook for lifting hotplate rings

4.6 CHIMNEY

Three levels of blocks are needed for the base of the chimney. Cut the blocks as shown in Figure 59 at ≈ 11 cm. This will give you the right cross sectional area for the entrance at the base of the chimney.

On the fourth level, cut a hole in the block for the ash hatch (Figure 60). This will make cleaning the chimney easier.

The gap at the chimney entrance, where the flue gases leave the stove and enter the base of the chimney, now measures 11 cm x 23 cm, providing a cross sectional area of 253 cm^2 (Figure 61).

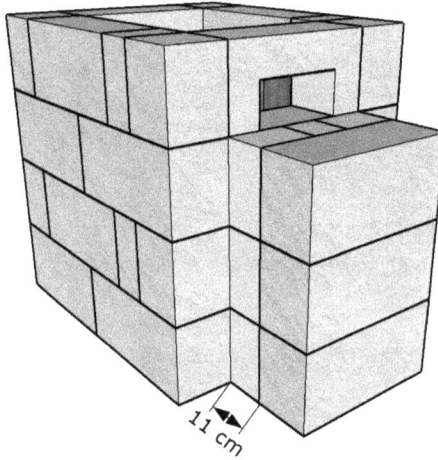

Figure 59: Base of the chimney

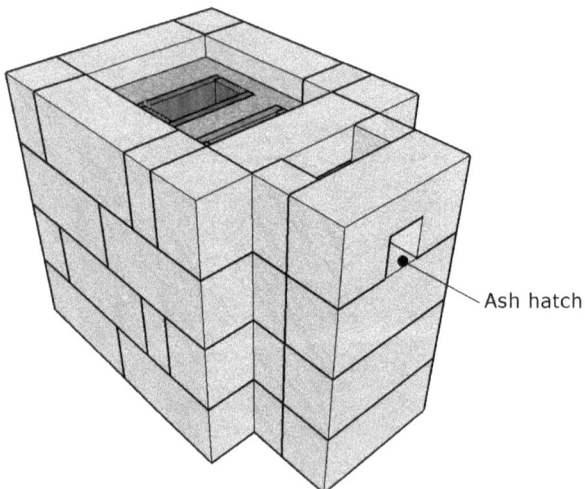

Ash hatch

Figure 60: Ash hatch hole

253 cm2

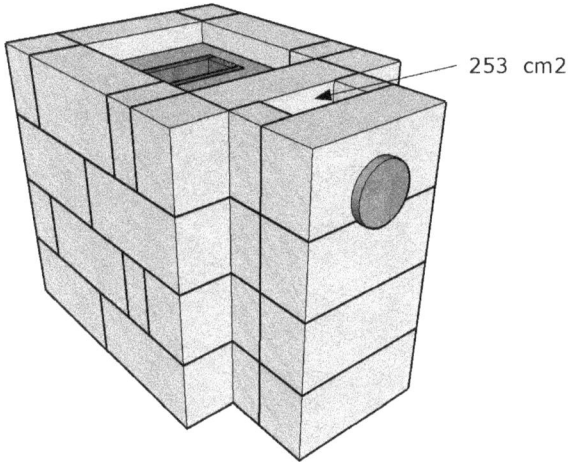

Figure 61: Cross sectional area at the chimney entrance

If you're using concrete chimney pipes, look for ones that have an internal diameter of ≈ 18 cm. This way, the chimney will have a cross sectional area of 255 cm² (Figure 62), almost identical to everywhere else in the stove. Use the following calculation:

255 cm2

➢ Area of a circle (**a**) = π * r²

Where:

➢ π = 3.142

➢ r = 9

18 cm

Therefore:

➢ **a** = 3.142 * 9²

➢ a = 3.142 * 81

➢ a = **255 cm²**

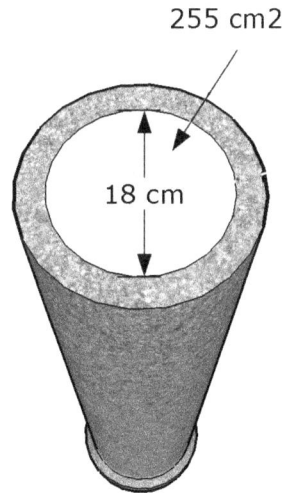

Figure 62: Cross sectional area of the chimney pipe

Figure 63: Final finish with white cement (courtesy of Silvestre Sales)

4.7 FINISH

The last remaining detail is the finish. This can be done with a white cement and water render, which gives the stove a clean aesthetic and seals the gaps around the hotplate, making sure the flue gases don't escape (Figure 63).

The finish can also be done with tiles, which gives an even better result and makes cleaning easier, while adding cost and extending construction time (Figure 64).

4.8 ALTERNATIVE DESIGNS

There are, of course, a wide variety of stove designs you can work with. The design presented here sacrifices some thermal efficiency at the expense of ease of use while getting smoke out of the kitchen. If you follow the design criteria explained above you can adapt the design to meet the needs of end-users and available materials in your area.

Figure 64: Finsih with tiles

Figure 66: AIDG metal stove, Xelateco, Quetzaltenango, Guatemala
(courtesy of Ben Dana, AIDG)

Figure 65 shows an example of an all metal stove on a mobile base, designed and built by Ben Dana and the AIDG in Quetzaltenango, Guatemala.

Another example is the ONIL "Plancha" stove developed by HELPS International (Figure 66), made from pre-fabricated concrete parts and cinder blocks.

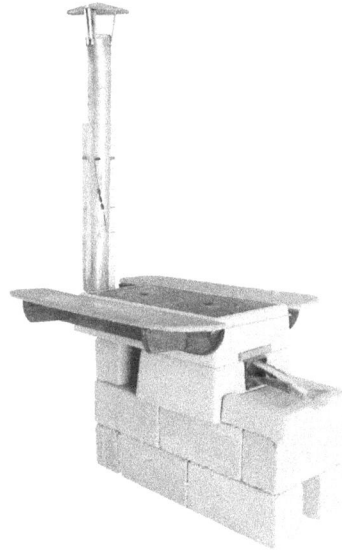

Figure 65: ONIL plate stove (courtesy of HELPS International)

CHIMNEY

POT

INSULATING
MATERIAL

OIL BARREL

FUEL

FIREBOX

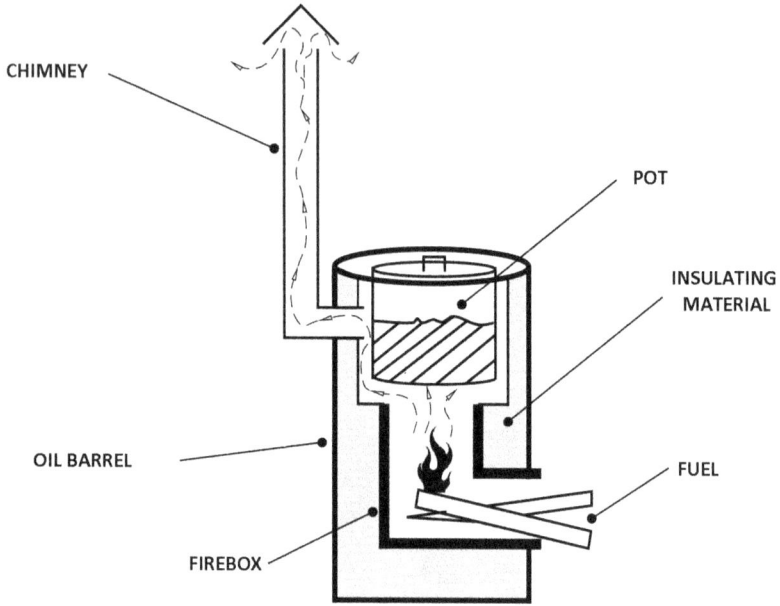

Figure 67: Rocket barrel stove

An alternative and, in principle, more efficient design (that also gets smoke out of the kitchen), is the rocket barrel stove (Figure 67).

The advantage of these stove designs is that they can be pre-fabricated and mass produced, which improves their quality, reduces their cost, and helps large scale roll-out. Whatever design you choose and however you decide to build, assemble and install the stoves, remember: stoves are for people, not for filling in reports, justifying project funding and talking big numbers. Involve the people who are going to use the stoves, evaluate stoves that are built and do a thermal efficiency test (described below) as a process of continual improvement.

5 Operation, maintenance, and thermal efficiency test

5.1 Operation and maintenance

The operation and maintenance of the Maya Petén stove is all fairly straightforward. However, for people used to cooking on open fires, the change can be significant and there can be frequent problems. Here are general pointers:

➢ **Cement curing:** end-users need to wait at least 8 days before using the stove, wetting it down daily with water to make sure the cement dries as slowly as possible. The slower it dries the less it will crack.

➢ **Thermal shocks:** the hotplate and firebox <u>must not get wet</u> when they are hot. This will give them a thermal shock and they will warp and break.

➢ **Pots and pans:** advise end-users to use metal pots and pans rather than clay or ceramics, as they conduct heat better and will result in lower fuel consumption.

➢ **Cleaning the chimney:** the chimney needs to be cleaned every 6 months, together with a visual inspection of the chimney pipes to make sure there are no signs of failure.

➢ **Hotplate:** clean with a damp cloth every morning <u>before cooking.</u>

➢ **Insulating material:** make sure end-users check the level of the insulating material beneath the hotplate every month to make sure it´s at the right level, topping up if necessary.

➢ **Fuel insertion:** make sure end-users are careful when they put logs into the firebox, so as not to break the tiles. Advise them that smaller logs are better.

5.2   THERMAL EFFICIENCY TEST

A thermal efficiency test is an important way to test the effectiveness of your design. The test presented here is a simplified version of the water boiling test developed by the University of California in Berkeley, based on the Vita International Standard Water Boiling Test presented by the Aprovecho Research Centre.

The test is based on boiling water and the measurement of the quantity of wood burnt and boiling time. The result provides an indication of the thermal efficiency of the stove. Figure 69 and Figure 70 contain the Data Sheet and the Worksheet for recording data and calculating the results. The test is divided into three parts:

1.   Cold start

2.   Hot start

3.   Simmer

The first test (Cold start) is done when the stove is cold, at least 24 hours after the last use. The second test (Hot start) is done immediately after the first test, followed by the third test (Simmer). The tests needs to be done in a sheltered area protected from wind, and require the following:

➤   30 kg of air-dried logs

➤   1 set of scales with a capacity ≥ 6 kg and 1 gram accuracy

➤   1 heat resistant pad to protect the scale

➤   1 digital thermometer, accurate to 0.1 ºC with 1 thermocouple probe that can be inserted into liquids

➤   1 timer

➤   1 pot (without a lid)

➤   1 small shovel or spatula and 1 dustpan for removing charcoal and ash

➤   1 tray for charcoal

➤   1 pair of fire resistant gloves

➤   3 bundles of wood:

  #1: Cold start: ≈ 2 kg

  #2: Hot start: ≈ 2 kg

  #3: Simmer: : ≈ 5 kg

5.2.1 Start of test

a. Measure the air temperature and note the result on the Data Sheet.

b. Weigh the pot, without a lid, and record the result.

c. Weigh the charcoal tray and record the result.

d. Prepare and label the 3 bundles of wood (2 kg / 2 kg / 5 kg). Use logs of approximately the same size and record their dimensions.

5.2.2 Test # 1: Cold Start

1. Fill the pot with 5 litres of water @ ≈ 20ºC, weigh the pot with water and record the result.

2. Insert the digital thermometer in the water, in the centre of the pot, 5 cm from the bottom, and record the temperature.

3. Light the fire with Bundle # 1, without using too much wood.

4. Once the fire is lit, start the timer.

5. When the water starts to boil:

    a. Note the time, measure the water temperature and record the result.

    b. Remove the wood and put out the flames. Remove the charcoal and logs.

    c. Weigh the unburnt wood you've removed from the stove and any logs leftover from the bundle.

    d. Weigh the pot with water and record the result.

    e. Remove the coals and charcoal from the stove and knock of the charcoal from the logs, weigh and record the result.

    f. Move on the Test # 2: Hot Start (do this quickly without letting the stove cool down)

5.2.3 Test # 2: Hot Start

1. Fill the pot with 5 litres of water, weight the pot with water and record the result. Measure the temperature of the water and record the result.

2. Light the fire with wood from Bundle # 2, without using an excessive amount of wood.

3. Once the fire is lit, start the timer.

4. When the water boils:

    a. Note the time, measure the water temperature and record the results.

    b. Remove the wood from the stove and put out the flames. Remove the charcoal from the logs.

    c. Weigh the unburnt wood from the stove and any remaining wood in the bundle and record the result.

    d. Weigh the pot with water and record the result.

    e. Remove the coals and charcoal from the stove, knock of the charcoal from the logs, weigh and record the result.

    f. Start immediately with Test # 3: Simmer, without letting the stove cool down.

Figure 68: A Maya Petén stove following 2 months of use, Intag, Ecuador

5.2.4   Test # 3: Simmer

1.   Fill the pot with 5 litres of water, weigh the pot with water and record the result. Measure the water temperature and record the result.

2.   Once the fire is lit, start the timer.

3.   Once the water boils, remove the pot from the stove and:

   a.   Note the time, measure the water temperature and record the results.

   b.   Put the pot back on the stove.

   c.   Weigh what´s left of the wood and record the result.

   d.   Put the thermometer in the water and adjust the amount of fire so that the water temperature remains, as far as possible, 3 ºC below boiling point.

1.   Measure the water temperature and record the result.

2.   Note the time, and for the following 45 minutes, adjust the fire so that the water temperature remains, as far as possible, 3 ºC below boiling point.

3.   Once 45 minutes have passed, note the time (45 minutes!), measure the water temperature, record the result, and:

   a.   Remove the wood from the stove and put of the flames. Remove charcoal from the logs.

   b.   Weigh the unburnt wood from the stove and anything remaining in the bundle, and record the result.

   c.   Weigh the pot with water and record the result.

   d.   Remove the coals and charcoal from the stove, knock of any charcoal from the logs, weigh and record the result.

Water temperatures for Test #3 can vary slightly, but it´s important to try and keep the water temperature as close as possible to 3 ºC below boiling point. If the temperature drops to more than 6 ºC below boiling point, then the test must be cancelled and repeated. Use short smaller logs to regulate the fire temperature.

For more accurate results, do the tests above three times on the same stove and take the average of each respective result.

5.2.5   Results analysis

Using the Work Sheet below:

➢   Calculate the time need to reach boiling point for each test.

➢   Calculate the weight of wood used in each test, subtracting the starting wood weight from the final wood weight.

➢   Calculate the amount of water lost in each test, subtracting the starting water weight from the final weight.

➢   Do the same for the charcoal that is produces.

➢   Use the results to calculate the thermal efficiency of the stove.

Figure 69 and Figure 70 below contain the Data Sheet and Work Sheet for data entry and for calculating the thermal efficiency of the stove. Stoves that make a lot of charcoal are not working efficiently. Remember that you can´t compare test results of stoves done in different places at different times.

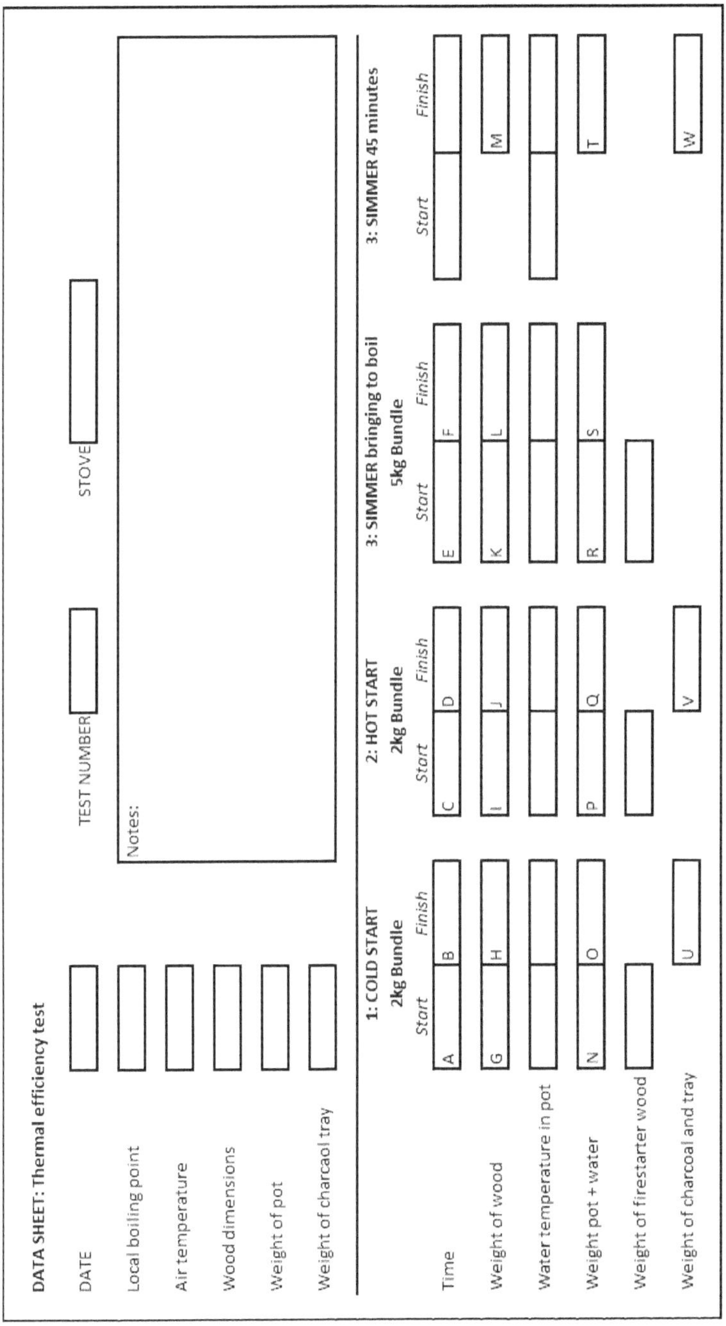

DATA SHEET: Thermal efficiency test

DATE

Local boiling point

Air temperature

Wood dimensions

Weight of pot

Weight of charcaol tray

TEST NUMBER

STOVE

Notes:

	1: COLD START 2kg Bundle		2: HOT START 2kg Bundle		3: SIMMER bringing to boil 5kg Bundle		3: SIMMER 45 minutes	
	Start	Finish	Start	Finish	Start	Finish	Start	Finish
Time	A	B	C	D	E	F		M
Weight of wood	G	H	I	J	K	L		
Water temperature in pot								
Weight pot + water	N	O	P	Q	R	S		T
Weight of firestarter wood								
Weight of charcoal and tray		U		V				W

Figure 69: Data sheet for thermal efficiency test

WORK SHEET: Thermal efficiency test

DATE [   ]   TEST NUMBER [   ]   STOVE [   ]

Time to boil

[   ] = B - A = Time to boil, Cold Start

[   ] = D - C = Time to boil, Hot Start

[   ] = F - E = Time to boil, Simmer

Wood used

[   ] = G - H = Wood use, Cold Start

[   ] = I - J = Wood use, Hot Start

[   ] = K - L = Wood use, Simmer, boiling phase

[   ] = L - M = Wood use, Simmer, 45 minutes

Water to steam

[   ] = N - O = Water to steam, Cold Start

[   ] = P - Q = Water to steam, Hot Start

[   ] = R - S = Water to steam, Simmer, boiling phase

[   ] = S - T = Water to steam, Simmer, 45 minutes

Charcaol created

[   ] = U - Y = Charcoal made, Cold Start

[   ] = V - Y = Charcoal made, Hot Start

[   ] = W - V = Charcoal made or consumed during Simmer
If this number is positive, then additional charcoal was created during simmer phase
If this number is negative, then charcoal was consumed during simmer phase

Figure 70: Worksheet for thermal efficiency test

6 BIBLIOGRAPHY

Baldwin S.F. (1986), *Biomass Stoves: Engineering Design, Development, and Dissemination*, Princeton University, VITA, Virgnia 22209 USA.

Bryden Dr. M., Still D., Scott P, Hoffa G., Ogle D., Bailis R., Goyer K. (2006), *Design Principles for Wood Burning Cook Stoves,* Aprovecho Research Center, Shell Foundation, Partnership for Clean Indoor Air. Cottage Grove, OR 97424, USA.

CIBSE (2006), *CIBSE Guide A: Environmental Design,* CIBSE Publications, United Kingdom.

Dana B. (2009), *Design manual Rocket Box Cook Stove*, AIDG Appropriate Infrastructure Development Group. Providence, RI 02909, USA.

FAO (1993), *Improved solid biomass burning cookstoves: a development manual*, Regional wood energy development programme in Asia GCP/RAS/154/NET. FAO Regional Wood Energy Development Programme in Asia, Bangkok, Thailand.

Scott P., *Una guía simple para construir la Estufa Justa,* Aprovecho Research Center, Shell Foundation, Partnership for Clean Indoor Air. Cottage Grove, OR 97424, USA.

7 ABOUT THE AUTHOR

Oliver Style worked happily for 10 years in Mexico, Colombia and Ecuador, getting bitten by mosquitoes and delivering projects and training programmes in off-grid photovoltaic systems, efficient biomass cooking stoves, and water supply systems, for the NGO Concern America. Based in Barcelona, Spain, he currently works as a consultant for Concern America's appropriate technology programmes, alongside work as a Passive House designer.

www.ingramcontent.com/pod-product-compliance
Lightning Source LLC
Chambersburg PA
CBHW050523210326
41520CB00012B/2414